The Development And Homologies Of The Mouth Parts Of Insects

Vernon Lyman Kellogg

THE

AMERICAN NATURALIST

VOL. XXXVI.   *September, 1902.*   No. 429.

THE DEVELOPMENT AND HOMOLOGIES OF THE MOUTH PARTS OF INSECTS.

VERNON L. KELLOGG.

THE problem of the homologies of the mouth parts of insects is one long worked at by zoölogists. Since Savigny's first statement in 1816 of his conclusions regarding the homologies of the arthropod appendages, this problem has been a favorite one with insect morphologists, and in this century of work much has been accomplished. There is a practically complete agreement as to the homologies of the parts of the biting mouth as this mouth is variously composed in the Orthoptera, Coleoptera, Neuroptera, *et al.*, and a fair agreement obtains with regard to the interpretation of the homologies of some of the more modified kinds of mouth parts possessed by the piercing and sucking insects. This is true especially of those insects, like the Hymenoptera and the Lepidoptera, among which there are generalized forms showing the essential biting type (as with the sawflies among the Hymenoptera and Eriocephala and Micropteryx among the Lepidoptera), together with a series of gradatory forms leading plainly up to the highly specialized conditions exhibited by the higher members of these orders.

These homology-determinations were first made by a study of the comparative anatomy of the fully developed mouth parts (those of adult insects), and indeed have a fairly safe grounding on this comparative anatomical study alone. But with the development of embryological studies of insects came the confirmation of these determinations, or some of them, by the study of the development of the mouth parts. From their origin as budding appendages, arising on the successive segments of the embryonic head, their development has been readily and certainly traced to the definitive mouth-part condition ; and mandibles, maxillæ, and labium are as certainly serially homologous with each other and with the legs and antennæ as are the more obviously homologous segmental appendages of the crustaceans.

But this ontogenic development of the insectean mouth parts, simple and continuous as it is in the case of insects with an incomplete metamorphosis, is a very complex and difficult subject of study in all of those insects which undergo what is termed a complete metamorphosis, and this for the reason, now familiar to entomologists, that in the late larval and early pupal life of such insects a more or less radical histolysis, or breaking down of the larval organs and tissues, occurs, with a building up of the imaginal organs from small, primitive centers called histoblasts (imaginal disks), which are not derived from the corresponding larval organs but (for the appendages as legs and mouth parts) from the larval derm or cellular skin layer. Thus we have in the development of the mouth parts of insects with complete metamorphosis a discontinuity which sadly interferes with the determination of homologies by ontogenetic study. Indeed; so serious has this obstacle proved that we have as yet practically no complete tracing through both embryonic and post-embryonic development of the growth and development of the mouth parts of any insect of complete metamorphosis. And they are, for the most part, precisely those insects of most radical post-embryonic metamorphosis which possess in adult condition the most highly modified and specialized mouth parts, and which present to us the most serious task in the interpretation of the mouth-part homologies. The Diptera, of course, best exemplify these conditions.

There is no special difficulty, outside of the general difficulties which the study of insect embryology commonly presents, in tracing from beginning up to completed larval condition the development of the mouth parts of insects with complete metamorphosis; and the homologies of these larval mouth parts with the mouth parts of adult insects with incomplete metamorphosis can accordingly be determined on a basis of ontogenic study (also, of course, on a basis of comparative anatomy). The biting mouth parts of the more generalized flies, of the lepidopterous caterpillars and coleopterous grubs, can be homologized with the mandibles, maxillæ, and labium of the adult cockroaches and locusts, constituting the generalized biting or so-called orthopterous mouth. But when the attempt is made to carry the homologies on to the adult piercing and sucking mouths of the flies and butterflies we lose in the prepupal stage our grip on the continuity of embryonic and adult mouth conditions and find ourselves forced to rest our interpretation of the homologies of the adult dipterous, lepidopterous, and hymenopterous mouth on the basis of comparative anatomical studies. And fortunately for us the persistence of certain generalized forms already referred to enables us to make a pretty secure determination of these homologies for all of the orders except the Diptera. To my mind, indeed, the study of the comparative anatomy of the mouth parts of the generalized flies (families of the Nematocera) enables us to be pretty certain even in that order, but such an attempt[1] of mine in 1899 has certainly failed to be convincing to several entomologists.

There is necessary, then, the completion of the tracing of the development of the mouth parts; nothing less, under the circumstances that the most generalized of dipterous mouths are not at all generalized (if one may be so paradoxical), but are so specialized that no safe determination of the homologies can be made on the basis of comparative anatomy, — nothing less will be convincing or satisfactory for the solid grounding of an interpretation of the homologies of the mouth parts of

---

[1] The Mouth Parts of the Nematocerous Diptera, *Psyche*, vol. viii (1899): I, pp. 303–306, January; II, pp. 327–330, March; III, pp. 346–348, April; IV, pp. 355–359, May; V, pp. 363–365, June; with 11 figs.

the Diptera, and if this tracing can be effected for the other orders of holometabolous insects, it will put the homology-determinations on a much better foundation than they now have. It is the beginnings of such an attempt that is outlined in this paper.

NEUROPTERA.

The Neuroptera belong to the holometabolous insects, *i.e.*, insects with complete metamorphosis, but this metamorphosis in many forms is of a very simple and straightforward kind as compared with the radical metamorphosis of a fly or butterfly,

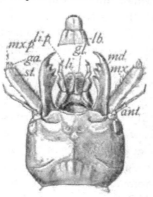

for example. The mouth parts of the adult insect are, too, of the orthopterous or biting type, and there is no question regarding the interpretation of the mouth-part homologies. Mandibles, maxillæ, and labium of the neuropterous mouth are obviously homologous with the similarly named parts of the orthopterous mouth. Furthermore, the differences between the larval and adult mouth parts are comparatively slight, and no question is made regarding the homologies between the two sets. Yet it is worth while to trace the development of the imaginal parts in its more conspicuous features, and get a first sight

FIG. 1.— Dorsal aspect of head of larva of *Corydalis cornuta*, with labrum removed. *lb.*, labrum; *md.*, mandible; *mx.*, maxilla; *mx.p.*, maxillary palpus; *st.*, stipes; *ga.*, galea; *li.*, labium; *li.p.*, labial palpus; *ant.*, antenna.

at the relation between larval and imaginal mouth parts in a holometabolous insect. This relation is readily made out in the large and familiar neuropteron called the "dobson fly," or "hellgrammite," *Corydalis cornuta.*

Corydalis cornuta (Figs. 1–5).—The mouth parts of the larval Corydalis are shown in Fig. 1, and their orthopterous character, together with the details of the various parts, are so readily apparent that little description is needed. The mandibles (*md.*) are very heavy and long ; the maxillæ (*mx.*) have a short proximal segment, cardo (not visible in the drawing), and a usually elongate parallel-sided stipes (*st.*) bearing at its terminal extremity the much-reduced three-segmented palpus (*mx.p.*) and a still smaller two-segmented terminal lobe, or galea (*ga.*), the lacinia being wholly wanting; the labium (*li.*) has the glossæ (*gl.*) (inner terminal lobes) fused but emarginate, the paraglossæ (outer terminal lobes) wanting, and the palpi (*li.p.*) three-segmented and well developed. The mouth parts are similar in both sexes.

When the larval dobson is ready to pupate (at the probable age of three years) it leaves the stream it has lived in, crawls under some stone near the water's edge, and changes into a quiet, non-feeding pupa, which, however, is not enclosed in a hard, opaque cuticle, but retains the power of violent wriggling, and bears the wing pads and legs only loosely appressed to the body. The mouth parts of the pupa (Fig. 2) show slight yet obvious differences from those of the larva (and also from those of the imago). The mandibles (*md.*) show a difference from the larval mandibles in the character of the dentation and in outline of the whole sclerite; the maxillæ have short, five-segmented palpi and two short terminal lobes, *i.e.*, both galea (*ga.*) and lacinia (*lc.*), and the labium (*li.*) has its free margin more emarginate and less truncate or blunt, the palpi (*li.p.*) remaining three-segmented.

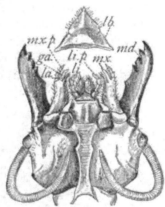

Fig. 2.—Ventral aspect of head of pupa of *Corydalis cornuta*. *lb.*, labrum; *md.*, mandible; *mx.*, maxilla; *mx.p.*, maxillary palpus; *ga.*, galea; *la.*, lacinia; *li.*, labium; *li.p.*, labial palpus.

Fig. 3.—Dorsal aspect of head of old larva of *Corydalis cornuta*, with body wall of right side (in figure) dissected away, showing pupal head beneath. *l.h.*, larval head wall; *l.md.*, larval mandible; *l.mx.*, larval maxilla; *l.li.*, larval labium; *l.ant.*, larval antenna; *p.h.*, wall of pupal head; *p.md.*, pupal mandible; *p.mx.*, pupal maxilla; *p.li.*, pupal labium; *p.ant.*, pupal antenna.

If one dissects away the cuticle of the head of an old larva about to pupate, the pupal mouth parts will be found formed fairly within the old larval ones, and thus in perfect correspondence with them. Rather it would be truer to say that they are apparently the transformed larval parts minus the to-be-shed larval cuticle. This is shown in Fig. 3, in which the larval cuticle of the right-hand half of the head (including the whole of the labium) has been dissected away, exposing the

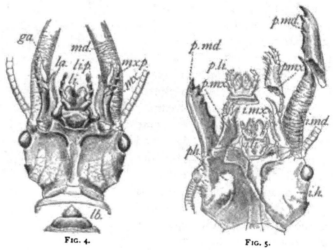

FIG. 4.

FIG. 5.

FIG. 4. — Ventral aspect of head of adult male *Corydalis cornuta*. *lb.*, labrum; *md.*, mandible; *mx.*, maxilla; *mx.p.*, maxillary palpus; *ga.*, galea; *la.*, lacinia; *li.*, labium; *li.p.*, labial palpus.

FIG. 5. — Ventral aspect of head of pupa of *Corydalis cornuta*, the pupal body wall being dissected away on right side (in figure), showing forming imaginal head and appendages. *p.md.*, pupal mandible; *p.mx.*, pupal maxilla; *p.li.*, pupal labium; *p.h.*, pupal body wall of head; *i.md.*, imaginal mandible; *i.mx.*, imaginal maxilla; *i.li.*, imaginal labium; *i.h.*, body wall of imaginal head.

still soft, unchitinized pupal cuticle, while the left side of the head is still wholly larval. From the right pupal mandible has been slipped the larval mandibular sheath, from the right pupal maxilla has been slipped the larval maxillar sheath, and from the whole pupal labium has been removed the larval covering. But the slight changes in outline and character of the pupal mouth parts are plainly apparent, while the identity of larval and pupal mandibles, maxillæ, and labium is unmistakable. There is yet no apparent difference in the mouth parts of the sexes.

In the adult (Fig. 4) we find mouth parts still of simple orthopterous type, with parts plainly homologous with the various orthopterous parts, and also as plainly with the parts

of its own larva and pupa; but in the male the familiar but extraordinary modification of the mandibles, converting them from biting and masticating organs into a pair

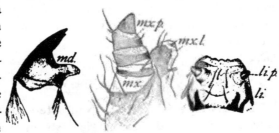

FIG. 6.—Mouth parts of larva of *Anatis 15-punctata*. *md.*, mandible; *mx.*, maxilla; *mx.l.*, maxillar lobe; *mx.p.*, maxillary palpus; *li.*, labium; *li.p.*, labial palpus.

of long, non-dentate, pointed, clasping organs (*md.*) for holding the female, attracts our special attention. But these organs are certainly mandibles; the maxillæ (*mx.*) and labium (*li.*), changed slightly to be still more thoroughly orthopterous in type, are in their own places, and no other mandibulate organs except the claspers are present. In the female the imaginal mandibles are of usual biting dentate type. To make sure of the mandibulate character of the long claspers we have but to dissect the head of an old pupa, as shown in Fig. 5. In

FIG. 7.—Mouth parts of adult *Anatis 15-punctata*. *md.*, mandible; *mx.*, maxilla; *mx.p.*, maxillary palpus; *ga.*, galea; *la.*, lacinia; *li.*, labium; *li.p.*, labial palpus.

this figure the pupal cuticle has been removed from the right-hand half of the head, while left intact on the left side. Removing the pupal labial cuticle, the imaginal labium, practically identical with the pupal one, is exposed, with the palpi shortened by "telescoping" but ready to expand to full length;

within the pupal maxillar sheath the imaginal maxilla in its now thoroughly orthopterous character is found, and within the comparatively short, strongly dentate, pupal mandible is found, strongly "telescoped," the strange adult mandible, with its lack of dentation, its pointed tip, and its great length (easily attained by extension of the longitudinally compressed organ as discovered within the pupal sheath). Thus the transformation of larval parts into pupal, and of pupal into imaginal, is obvious, and the homologies between larval and imaginal parts are firmly founded on ontogenic basis.

COLEOPTERA.

The Coleoptera, like the Neuroptera, have biting mouth parts in both larval and imaginal stages, but the differences are usually greater, and the general metamorphosis is on the whole more radical.

Anatis 15-punctata (Figs. 6–8). — The accompanying figures made from a study of the mouth parts of *Anatis 15-punctata* illustrate the relations between larval and imaginal mouth parts of a member of the order. The larvæ (Fig. 6) have strongly chitinized, sharp-toothed mandibles (*md.*), maxillæ (*mx.*) with single terminal lobe (*mx.l.*), rather large four-segmented palpus (*mx.p.*), and fleshy liplike labium (*li.*), with fused terminal lobes and short one-segmented palpus (*li.p.*) inserted on a segment-like projection. In the adult (Fig. 7) the mandibles (*md.*) are shorter and heavier, the maxillæ (*mx.*) have both terminal lobes, galea (*ga.*) and lacinia (*lc.*), distinct, and four-segmented palpi (*mx.p.*), the distal segment being much broader than the others. The labium (*li.*) is rather elongate, with distinct basal sclerites (submentum and mentum), fused terminal lobes, and short three-segmented palpi (*li.p.*).

The small size of the larval head precludes such dissections as were easily made in the case of Corydalis, and the thickness and opacity of the chitinized cuticle of the head makes it impossible to clear specimens and study the forming imaginal head within, a method very successfully used in the cases of the honeybee and digger wasp (see *postea*). The development of the imaginal head and mouth parts had to be studied by means of sections, and here again the firmness of the head wall offered a serious obstacle to satisfactory work. I have been able, however, to get series showing plainly the later steps of the development of the imaginal parts within the head of old larvæ. The developing imaginal parts, their definitive outlines already so strongly indicated as to make them recognizable (apart from their position), lie within the corresponding parts of the larval head (Fig. 8), imaginal mandibles with their tips within the larval mandibles, imaginal maxillæ with their two terminal lobes lying partly within and corresponding to the single terminal lobe of the larva, and imaginal palpi lying almost wholly within the larval palpi, and finally imaginal labium lying in the base of the larval labium. All of the forming imaginal parts are plainly seen to be folds or evaginations of the forming imaginal derm layer, which shows in sections as a continuous broad cellular line lying just underneath the larval integument.

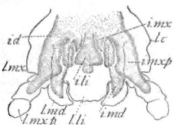

Fig. 8. — Semi-diagrammatic drawing of anterior portion of frontal horizontal section through the head of old larva of *Anatis 15-punctata* showing pupal (= imaginal) mouth parts forming underneath the larval integument. *l.c.,* larval cuticle; *i.d.,* imaginal derm; *l.md.,* larval mandible; *i.md.,* imaginal mandible; *l.mx.,* larval maxilla; *l.mx.p.,* larval maxillary palpus; *i.mx.,* imaginal maxilla; *i.mx.p.,* imaginal maxillary palpus; *l.li.,* larval labium; *i.li.,* imaginal labium.

Thus in Anatis we have practically the same conditions of development of the imaginal mouth parts within, and corresponding to, the larval mouth parts as we found in Corydalis.

Lepidoptera.

Among the Lepidoptera we find a great range in degree of specialization of the mouth parts. In Eriocephala and Micropteryx, as described by Walter [1] and myself,[2] the mouth parts are really of the biting type, the mandibles being short, heavy, and dentate, true jaws, the maxillæ showing a cardo, stipes, short galea, and lacinia, and long six-segmented palpus, and the labium being liplike, with plainly distinguishable submentum

[1] Walter, A. Beiträge zur Morphologie der Schmetterlinge, *Jenaische Zeitschr. f. Naturwiss.,* vol. xviii (1885), pp. 751–807.

[2] Kellogg, V. L. The Mouth Parts of the Lepidoptera, *Amer. Nat.,* vol. v (1895), pp. 546–556, Pl. XXV.

and mentum and prominent three-segmented palpi. But in all the Lepidoptera above the Eriocephalidæ, Micropterygidæ, and Tineidæ, from a considerable to a very pronounced speciali-

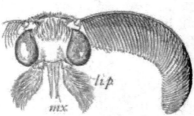

FIG. 9. — Frontal aspect of head of imago of *Notolophus leucostigma. mx.,* maxilla; *li.p.,* labial palpus.

zation is present, manifested by a complete reduction of the mandibles, by the reduction of the labium to a small rigid plate on the ventral side of the mouth bearing the persisting three-segmented palpi, and by a remarkable modification of the maxillæ whereby the galeæ (or laciniæ) are prolonged, grooved on their inner surfaces, and apposed to form the familiar sucking proboscis, while the other parts of the maxillæ are reduced and fused to form a rigid supporting base for the proboscis. In numerous moths no food is taken in the adult condition, and here the proboscis itself is reduced slightly or much, even to complete atrophy, and in extreme cases there is no mouth opening at all.

Notolophus leucostigma (Figs. 9–12). — In the white-marked tussock moth, *Notolophus leucostigma,* the mouth parts (Fig. 9) of the adult, although functionless, or at least apparently incapable of taking food, show all the usual parts peculiar to the typical specialized lepidopterous mouth. The labium is a small fixed plate, forming part of the ventral wall of the head and bearing the conspicuous hairy three-segmented palpi (*li.p.*); the maxillæ (*mx.*) are simply two slender

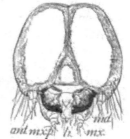

FIG. 10. — Frontal aspect of head of larva of *Notolophus leucostigma. ant.,* antenna; *md.,* mandible; *mx.,* maxilla; *mx.p.,* maxillary palpus; *li.,* labium.

tapering processes, the halves of the usual proboscis, but in this case not applied to each other and hence not forming a sucking tube ; the maxillary palpi are wholly reduced, and the mandibles entirely wanting.

In the caterpillar (Fig. 10) the biting mouth parts common to lepidopterous larvæ are present, with full complement of distinct and readily recognizable mandibles (*md.*), maxillæ (*mx.*) with short but distinct three-segmented palpi (*mx.p.*), and labium (*li.*) with very small ex-articulate palpi.

If an old larva, nearly ready to pupate, be taken, and its head dissected, as illustrated in Fig. 11, it will be found that

underneath, or within, the larval labium, or labial cuticle, will be found the forming imaginal labial palpi ; within the larval maxilla will be found the forming imaginal maxillæ, while within the larval mandible will be found nothing at all. In Fig. 11 the larval cuticle of the left side of the head has been dissected away, showing this correspondence between larval and imaginal parts ; the larval maxillary sheath has been slipped off of the forming imaginal maxillar process, while on that part of the forming imaginal head from which the larval mandible was taken there is not a trace even of a forming organ.   Fig. 12 shows the entirely dissected-out

FIG. 11. — Frontal aspect of head of old larva of *Notolophus leucostigma*, with body wall of left side (in figure) dissected away, showing p u p a l (= imaginal) head underneath. *l.ant.*, larval antenna ; *l.md.*, larval mandible ; *l.mx.*, larval maxilla ; *l.li.*, larval labium ; *lb.*, larval labrum ; *ant.*, imaginal antenna ; *i.mx.*, imaginal maxilla.

pupal (equals subimaginal head), with the already unmistakably recognizable imaginal mouth parts.

Thus in this representative of the Lepidoptera we find the imaginal mouth parts developing in perfect correspondence with the larval parts, imaginal maxillæ within larval maxillæ, imaginal labium in larval labium, and within the

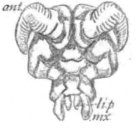

FIG. 12. — Pupal (= imaginal) head of *Notolophus leucostigma* dissected out of larval head.   *mx.*, maxilla ; *li.p.*, labial palpus ; *ant.*, antenna.

well-developed larval mandibles *nothing*, with a corresponding total absence of mandibles in the fully developed moth.   By sectioning the heads of old larvæ, it is readily perceivable that these developing imaginal mouth parts lying within and corresponding to the various larval parts are evaginations of the new or imaginal derm which forms a continuous layer underneath the larval integument. Similarly, it is apparent that the imaginal antennæ and compound eyes are in the one case evaginations and in the other simply modified portions of this imaginal derm; and although I have not made cuttings of a complete series of heads from young to oldest larvæ, enough of the younger stages have been studied to show the simple dermal origin of all these parts by a continuous process of evagination and modification. We are sufficiently acquainted with the origin and mode of development of the legs and wings of insects from histoblasts to recognize in these histoblasts, or developmental centers, simple invaginations of the derm, which later become evaginations. Whether an organ, as wing, leg, antenna, or mouth part, shall begin as an invagination or an evagination of the derm is chiefly a matter of mechanical necessity or ease, and of degree of radicalness in the metamorphosis. In either case the ultimate origin, that of being simply a particular portion or area of derm, is the same; the invagination must become an evagination; the difference lies in the mechanical factors of the developmental process.

HYMENOPTERA.

In the order Hymenoptera there is to be found, as in the Lepidoptera, a wide range of degree of specialization of the mouth parts, varying from the biting, orthopterous mouth of the sawflies to the highly modified sucking mouth of the honeybee; but throughout the order the mandibles persist in plainly jawlike character, and are always recognizable landmarks in mouth-part dissections. The only questions in the homology-interpretation occur in those cases where the labium and maxillæ are much modified and more or less completely fused or bound together. But these questions are not very serious; entomologists are fairly agreed, on a basis of comparative anatomical study, on the interpretation of the homologies of the hymenopterous mouth parts. But the results of a study of the post-embryonic development of the mouth parts, *i.e.*, the development of the imaginal mouth parts, undertaken by one of my students, Mr. M. H. Spaulding, illuminate too beautifully and effectively the whole study of the development of imaginal mouth parts in holometabolous insects to be overlooked because of the lack of any crying need for an ontogenic confirmation of the hymenopterous homologies. Mr. Spaulding has been admirably successful in so clearing and staining the heads of variously aged larvæ of the honeybee and of a digger wasp, Ammophila *sp.*, that the developing imaginal head within the larval integument may be as easily studied as the exterior of the larval head itself. The bee and wasp larvæ, it will be recalled, are both "inside feeders," *i.e.*, lie during their life enclosed in a protecting cell, in one case of wax, in the other of hardened mud, and thus may and do dispense with the heavily chitinized opaque head cuticle common to exposed insect larvæ. And both larvæ have full complements of mouth parts, namely, mandibles, maxillæ, and labium, — a condition not common to all larvæ in those two orders, Hymenoptera and Diptera, in which the post-embryonic metamorphosis is most radical. This condition is a necessary one for the determination of the relations of the imaginal to the larval parts.

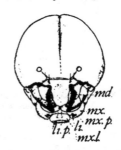

FIG. 13. — Frontal aspect of head of larva of digger wasp, *Ammophila* sp. *md.*, mandible; *mx.*, maxilla; *mx.p.*, maxillary palpus; *mx.l.*, maxillary lobe; *li.*, labium; *li.p.*, labial palpus.

Ammophila sp. (Figs. 13–15). — The larval mouth parts (Fig. 13) consist of well chitinized crushing mandibles (*md.*), short fleshy maxillæ (*mx.*) with very small one-segmented palpus (*mx.p.*) and smaller terminal lobe (*mx.l.*), and short liplike labium (*li.*) with pair of very small one-segmented palpi (*li.p.*). The adult wasp also has a complete complement of mouth parts (Fig. 14), all very elongate and slender, the mandibles (*md.*) heavily chitinized and toothed, the maxillæ (*mx.*) long and slender with distinct cardo and stipes, five-segmented palpus (*mx.p.*), and simple terminal lobe composed of the fused galea and lacinia, and the labium (*li.*) also long and narrow with fused submentum and mentum, four-segmented palpi (*li.p.*), slender ligula formed of the fused glossæ (*gl.*), and distinct slender paraglossæ (*p.g.*) less than half as long as the fused glossæ.

It is obvious that these long slender imaginal mouth parts cannot be contained within the very much shorter and altogether smaller larval parts. As a matter of fact, the whole imaginal head is for simple mechanical reasons forced to lie during its development chiefly in the anterior larval thoracic segment, the anterior portions, including the antennæ and mouth parts, projecting forward into the larval head capsule. But still there is indicated perfectly the correspondence between particular imaginal parts and particular larval parts

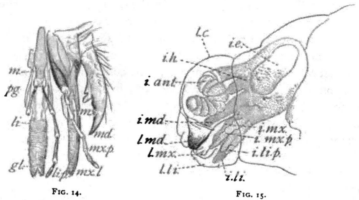

FIG. 14.   FIG. 15.

FIG. 14. — Mouth parts of adult digger wasp, *Ammophila* sp. (mandible, maxilla, and labial palpus of left side, in figure, not drawn). *md.*, mandible; *mx.*, maxilla; *mx.p.*, maxillary palpus; *mx.l.*, maxillary lobe; *li.*, labium; *li.p.*, labial palpus; *gl.*, glossæ; *pg.*, paraglossa; *m.*, mentum.

FIG. 15. — Head of old larva of digger wasp, *Ammophila* sp., cleared to show forming imaginal head within. *l.c.*, larval head wall; *i.h.*, forming imaginal head; *i.e.*, imaginal eye; *i.ant.*, imaginal antenna; *l.md.*, larval mandible; *i.md.*, imaginal mandible; *l.mx.*, larval maxilla; *i.mx.*, imaginal maxilla; *i.mx.p.*, imaginal maxillary palpus; *l.li.*, larval labium; *i.li.*, imaginal labium; *i.li.p.*, imaginal labial palpus.

by the fact that the projecting tips of the elongate imaginal parts penetrate or lie within the short larval parts. This is shown clearly in the cleared and stained heads prepared by Mr. Spaulding, as well as in series of sections. Fig. 15 is drawn with camera lucida from one of the whole head preparations, and, as indicated by the lettering, those parts of the imago which we have, on the basis of comparative anatomy, assumed to compose the labium, do project into and correspond with the larval labium; the case is similar with maxillæ and mandibles. But, in origin, these imaginal mouth parts arise as dermal modifications and outgrowths which for simple demands of space become far removed from the larval mouth parts, the bases of the developing imaginal parts lying, indeed, in late

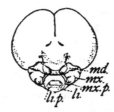

FIG. 16. — Frontal aspect of head of larva of honeybee, *Apis mellifica. md.,* mandible; *mx.,* maxilla; *mx.p.,* maxillary palpus; *li.,* labium; *li.p.,* labial palpus.

larval life in the first thoracic larval segment. But in earlier larval life the beginning imaginal parts lie almost wholly within the larval parts, and no one studying the series of whole head preparations and of sections can fail to be convinced of the certain correspondence and homology between larval and imaginal parts, although there may be said to be no perfect transformation or development of the one into the other, the evidence being that of a correspondence in position on the head and of part for part.

Apis mellifica (Figs. 16–18). — The beautiful series of cleared and stained heads of honeybee larvæ of different ages, and the series of sections of similar heads prepared by Mr. Spaulding, show a condition in the development of the imaginal mouth parts of the bee wholly identical with that just shown for the digger wasp. The larval mouth parts (Fig. 16) are very weakly chitinized, but are complete and readily distinguishable. They resemble in general the mouth parts of the digger wasp larva, but are smaller, weaker, and the short fleshy maxilla bears only the minute one-segmented palpus, having no tiny lobe as in the wasp maxilla. The imaginal mouth parts (Fig. 17) of the bee, familiar to all entomologists, are composed of horny, trowel-like mandibles (*md.*), long maxillæ (*mx.*) with cardo (*cd.*), stipes (*st.*), small one-segmented palpus

FIG. 17.— Mouth parts of adult honeybee, *Apis mellifica. md.,* mandible; *mx.,* maxilla; *mx.p.,* maxillary palpus; *mx.l.,* maxillary lobe; *st.,* stipes; *cd.,* cardo; *li.,* labium; *s.m.,* submentum; *m.,* mentum; *pg.,* paraglossa; *gl.,* glossa; *li.p.,* labial palpus.

(*mx.p.*), and with galea and lacinia fused to form a single flattened, pointed, bladelike terminal lobe (*mx.l.*), and of labium (*li.*) with long, tapering subcylindrical ligula formed of the fused glossæ (*gl.*), short but distinct flaplike paraglossæ (*pg.*), three-segmented palpi (*li.p.*) borne on a long palpiger, and at the base a distinct mentum (*m.*) and submentum (*sm.*).

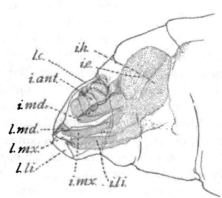

Fig. 18. — Head of old larva of honeybee, *Apis mellifica*, cleared to show forming imaginal head within. *l.c.*, larval head wall; *i.h.*, imaginal head; *i.e.*, imaginal eye; *i.ant.*, imaginal antenna; *l.md.*, larval mandible; *i.md.*, imaginal mandible; *l.mx.*, larval maxilla; *i.mx.*, imaginal maxilla; *l.li.*, larval labium; *i.li.*, imaginal labium.

As with the digger wasp, the developing head of the imago, with its long antennæ and mouth parts, demands more space than is afforded within the larval head segment, so that it is crowded backward and occupies part of the first and second larval thoracic segments. But the forming imaginal mouth parts are to be found with their tips projecting into the corresponding larval parts, as shown in Fig. 18. The conditions of the development of the imaginal parts, and of their perfect correspondence with the larval parts, are wholly like those already explained for the digger wasp.

DIPTERA.

In the case of the Diptera, — and it is here that the necessity of ontogenetic study is most important, indispensable indeed, for the determination of the homologies, — we have, as in the Hymenoptera and Lepidoptera, a great variety of mouth-part conditions culminating in the extreme specialization characteristic of the muscid forms. In most Diptera it is obvious that a total reduction of at least one pair of the buccal appendages has occurred, with a large reduction and complete modification of the remaining parts. From a considerable study of the anatomy of the fully developed mouth parts in a long series of dipterous forms, including representatives of all except one (the Ornephilidæ) of the nematocerous families, —

those families by common agreement held to constitute the
more generalized portion of the order, — I came to the conclu-
sion that the old and most widely, if perhaps uncritically,
accepted interpretation of the homologies of the dipterous one
is the true one.   This interpretation homologizes the labella-
bearing proboscis common to all the more specialized flies with
the labium of other insects, finds the maxillæ represented in
these specialized forms chiefly or only by a pair of palpi, and
finds the mandibles wholly wanting in all but the females of a
few families.   In the case of most of the nematocerous families
the labium retains a truly labiumlike character and has not
developed the pseudotracheæ-bearing labella, while the maxillæ are represented by a well-developed bladelike terminal
lobe as well as by the palpi.   The mandibles when present are
of the character of elongate blades or stylets, never of the
character of true crushing or biting jaws.   The structural
character of the mouth in each of the nematocerous families
is described and illustrated in my series of papers (1899) in
Psyche, previously referred to.

But several interpretations of the homologies of the mouth
parts widely at variance with the above have been offered.   In
these various interpretations the possession of mandibles by
any flies at all is denied; the so-called labium is considered to
be composed of modified parts of the maxillæ, and the so-called
maxillæ are believed to be parts of the labium; in fact, most of
the possible changes which an active speculation could invent
have been rung on the theme.   Nor are these interpretations
based on mere speculation; they are the results, in several cases,
of prolonged and disinterested examination of considerable
series of specimens.

In the face of such differences of opinion, and with the
apparent limits of the method of the comparative study of the
fully developed mouth parts of various members of the order
reached, it becomes imperative to seek the clue to these lost
homologies in the facts of development.   And this is really the
first object of this present study.   Can the homologies of the
dipterous mouth parts be discovered by the study of the development of the parts?

For a complete developmental study of the mouth parts of any dipteron it would be necessary to begin with the budding appendages of the head segments in early embryonic life, to trace the development of these appendages to their definitive form in the hatched larva, and finally to follow the transformation, if it occurs, of these larval parts into the ultimate imaginal ones. As a matter of fact, such actual transformation does not occur, so that the study of the postembryonic development of the mouth parts consists of noting the ecdysis of the larval parts and determining the ontogenic relations of the new imaginal parts to the old larval ones.

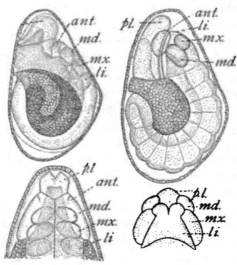

As for the embryonic development of the mouth parts,—*i.e.,* the development from budding appendages to definitive larval parts, — that has been done for several Diptera, and in particular by Metschnikov for Simulium, one of the two flies whose postembryonic development I

FIG. 19. — Two embryonic stages in the development of *Simulium* sp. (after Metschnikov): younger stage at left, older at right; in upper row whole embryos from lateral aspect, in lower frontal aspect of heads of same stages. *p.l.,* pro-cephalic lobes; *ant.,* antenna; *md.,* mandible; *mx.,* maxilla; *li.,* labium.

shall describe. These embryonic studies make certain the homologies of the larval parts; in those flies like Simulium, whose larvæ are provided with a biting mouth with full complement of parts, it is easy to note plainly the development of mandibles, maxillæ, and labium from the successive pairs of budding head appendages (Fig. 19), and thus to homologize these parts certainly with the mandibles, maxillæ, and labium of adult insects of incomplete metamorphosis. There remains to determine the relations of the larval mouth parts of Simulium with its very different imaginal mouth parts.

In selecting flies for the study of the postembryonic development of the mouth parts I have chosen two which in the imaginal condition possess all the parts possessed by any fly,

FIG. 20.— Mouth parts of adult *Simulium* sp., female. *l.ep.*, labrum-epipharynx; *hyp.*, hypopharynx; *md.*, mandible; *mx.*, maxilla; *mx.l.*, maxillary lobe; *mx.p.*, maxillary palpus; *li.*, labium; *pg.*, paraglossa.

and these parts in as generalized condition as is to be found in the order, and which also possess in the larval stage a similarly full complement of mouth parts. Such larvæ as those of the Muscidæ, with their problematical hooks and lack of other parts, and such imagines as the muscid flies, with no parts left except proboscis and maxillary palpi, are impossible for the determination of the relation between larval and imaginal parts. From the mouth parts of the imaginal Simulium and of other nematocerous forms it is not difficult to trace the evolution to the specialized muscid conditions, and if the mouth parts of Simulium and similarly equipped flies can be interpreted, the various members of the dipterous series culminating in the muscids can. So in Simulium and Blepharocera I have found suitable forms for study; both with females possessing the so-called mandibles, both with maxillæ and labium well developed in both sexes, and both with larvæ equipped with biting mouths with unmistakable mandibles, maxillæ, and labia, and in one case, that of Simulium, with the embryonic

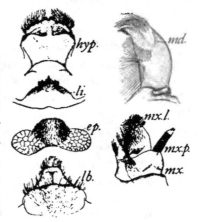

FIG. 21.— Mouth parts of larva of *Simulium* sp. *md.*, mandible; *mx.*, maxilla; *mx.l.*, maxillary lobe; *mx.p.*, maxillary palpus; *lb.*, labrum; *ep.*, epipharynx; *li.*, labium : *hyp.*, hypopharynx.

development of the larval mouth parts fully traced and the homologies certainly[1] determined.

[1] Metschnikov, E. Embryologische Studien an Insekten, *Zeitschr. f. wiss. Zool.*,

Simulium sp. (Figs. 20–23). — In the female imago[1] the mouth parts (Fig. 20) consist of a short liplike labium (*li.*) composed of a short basal sclerite and three terminal lobes, being the two large paraglossæ (*pg.*) and a median short membranous lobe, the fused glossæ ; of a pair of maxillæ (*mx.*), each consisting of a basal sclerite, a long five-segmented palpus (*mx.p.*), and a single pointed, bladelike terminal lobe (*mx.l.*) reaching nearly to the end of the third palpar segment, serrate on its inner margin at the tip and better developed than in most Nematocera ; and of a pair of short mandibles (*md.*), broad, thin, and weakly chitinized. As in other nematocerous flies, there is a well-developed labrum-epipharynx (*l.ep.*) and an elongate flattened hypopharynx (*hyp.*). In the males the mandibles are wanting.

In the larva (Fig. 21) the mouth is of the biting type, with short-toothed and heavy mandibles (*md.*), short, jawlike maxilla (*mx.*) with distinct one-segmented palpus (*mx.p.*), and a small, strongly chitinized labium (*li.*) or labial plate. In addition, labrum (*lb.*), epipharynx (*ep.*), and hypopharynx (*hyp.*) are all well developed.

The head of the larva having a thoroughly opaque, strongly chitinized cuticle, it was impossible to clear whole heads sufficiently to make visible the developing imaginal head and its parts, so that the method of sections had to be relied on to reveal the internal conditions. These sections of heads of larvæ of various ages show plainly that the general method of development of the imaginal parts within the larval head, and the correspondence between forming imaginal parts and the corresponding larval parts already noted in the other orders of holometabolous insects, hold good in the Diptera. Fig. 22 shows in sagittal longitudinal section the forming imaginal head parts within the larval head. This section shows particularly well the relation of the forming imaginal antenna to the

vol. xvi, 1866 ; embryonic development of mouth parts of Simulium described on pp. 392–421.

[1] In describing the adult mouth I shall assign to the various parts those names which, from my earlier study of the comparative anatomy, seem correctly used, and the use of which is confirmed by the results of this ontogenetic study.

larval antenna.   In the larva the antennæ are very small compared with their size in the imago, and the imaginal antenna is
thus forced, in its development, to occupy a region in the larval
head not included in the larval antenna.   But the tip of the
imaginal organ lies fairly within the larval organ, thus indi-
cating by correspondence in position, what is plainly obvious
from anatomical consideration, the homology between the larval
and imaginal organs.   Similarly the forming imaginal mouth
parts are to be found in unmistakable correspondence or
homologous relation with the larval parts.   By tracing the
development of the parts, marked in Fig. 22 as the forming

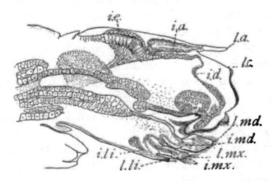

FIG. 22. — Sagittal section through head of old larva of *Simulium* sp., showing forming imaginal
head parts within.   *l.c.*, larval head wall; *i.d.*, imaginal derm; *l.a.*, larval antenna; *i.a.*,
imaginal antenna; *i.e.*, imaginal eye; *l.md.*, larval mandible; *i.md.*, imaginal mandible;
l.mx., larval maxilla; *i.mx.*, imaginal maxilla; *l.li.*, larval labium; *i.li.*, imaginal labium.

imaginal mouth parts, through larvæ of successively older ages
to pupation and the achievement of the definitive imaginal condition of these parts, it is certain that the parts marked respectively imaginal mandible, imaginal maxillæ, and imaginal labium,
lying respectively in the larval mandibles, maxillæ, and labium
(with homologies firmly based on ontogenic basis), do develop
into those definitive imaginal parts named mandibles, maxillæ,
and labium in Fig. 20, illustrating a dissection of the mouth
parts in a female adult Simulium.   Fig. 23, a horizontal, frontal
section through the head of a Simulium larva, shows also the
forming imaginal maxilla and mandibles within corresponding
larval parts.

Bibiocephala doanei [1] Kellogg (Figs. 24–26). — The Blepharoceridæ, or net-winged midges, agree with the Simulidæ, or black flies, in having the females equipped with mandibles, which in the Blepharoceridæ are well developed as long, slender, bladelike saws (see Fig. 24, *md.*) used to lacerate the bodies (as I have observed) of the tiny midges caught as prey by the bloodthirsty females. In addition the adult females have maxillæ (Fig. 24, *mx.*) with well-developed lobe (*mx.l.*) and long five-segmented palpus (*mx.p.*), and a labium (*li.*) consisting of strong elongate basal sclerite which presents indications of a line of fusion of submentum and mentum, and a pair of free fleshy terminal lobes, the paraglossæ (*pg.*). The males are equipped like the females except for the mandibles.

The larva of *Bibiocephala doanei* has a biting mouth (Fig. 25) composed of short, stout, crushing mandibles (*md.*), weaker jawlike maxillæ (*mx.*) without palpi, and a soft liplike labium (*li.*). In addition there are well-developed labrum-epipharynx (*l.ep.*) and hypopharynx (*hyp.*).

The development of the imaginal head shows the same phenomena as in Simulium. In Fig. 26, from a vertical transverse section through the head of an old larva, the derm of the forming imaginal head is plainly seen in continuous layer, modified at *i.e.* to produce the developing compound eyes and at *mx.* and *md.* the forming imaginal mandibles and maxillæ. In this section the imaginal parts of the maxillæ visible are the forming palpi, and their definitive, long, segmented condition is plainly to be seen in these telescoped organs tucked tightly inside the larval maxillæ. The forming mandibles do not yet show their definitive character, but in tracing these organs through a series of older larvæ the gradual taking-on of the slender sawlike character is manifest. The series of Blepharocera preparations which I have show even more plainly than the Simulium preparations the perfect correspondence and "box-in-box" sort of relation which exists between the larval

---

[1] This blepharocerid fly was described by me in *Psyche*, vol. ix (April, 1900), pp. 39–41, 2 figs., under the name *Liponeura doanei*. In a recent revision of the North American Blepharoceridæ, now in press, I refer this species to the genus Bibiocephala.

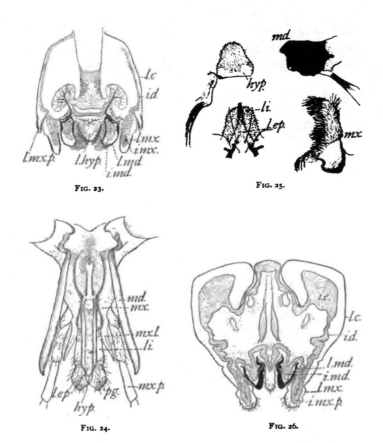

FIG. 23.

FIG. 25.

FIG. 24.

FIG. 26.

FIG. 23. — Frontal section, through the head of old larva of *Simulium* sp., showing forming imaginal parts. *l.c.*, larval cuticle; *i.d.*, imaginal derm; *l.md.*, larval mandible; *i.md.*, imaginal mandible; *l.mx.*, larval maxilla; *i.mx.*, imaginal maxilla; *l.mx.p.*, larval maxillary palpus; *l.hyp.*, larval hypopharynx.

FIG. 24. — Mouth parts of adult *Bibiocephala doanei*, female; *md.*, mandible; *mx.*, maxilla; *mx.l.*, maxillary lobe; *mx.p.*, maxillary palpus; *li.*, labium; *pg.*, paraglossa; *l.ep.*, labrum-epipharynx; *hyp.*, hypopharynx.

FIG. 25. — Mouth parts of larva of *Bibiocephala doanei*. *md.*, mandible; *mx.*, maxilla; *li.*, labium; *l.ep.*, labrum-epipharynx; *hyp.*, hypopharynx.

F:G. 26. — Frontal section, through head of old larva of *Bibiocephala doanei*, showing forming imaginal head parts within. *l.c.*, larval head wall; *i.d.*, imaginal derm; *i.e.*, imaginal eye; *l.md.*, larval mandible; *i.md.*, imaginal mandible; *l.mx.*, larval maxilla; *i.mx.p.*, imaginal maxillary palpus.

mouth parts, of whose homologies no doubt can exist, and the forming imaginal parts, of whose homologies, in definitive condition, I thought myself long ago able to speak confidently on a basis of comparative anatomical study, but of which now on a basis of ontogenetic study I am simply without doubt.

STANFORD UNIVERSITY, CALIFORNIA.
January, 1902.